This book belongs to

_____

_____

wonderful world
Book series

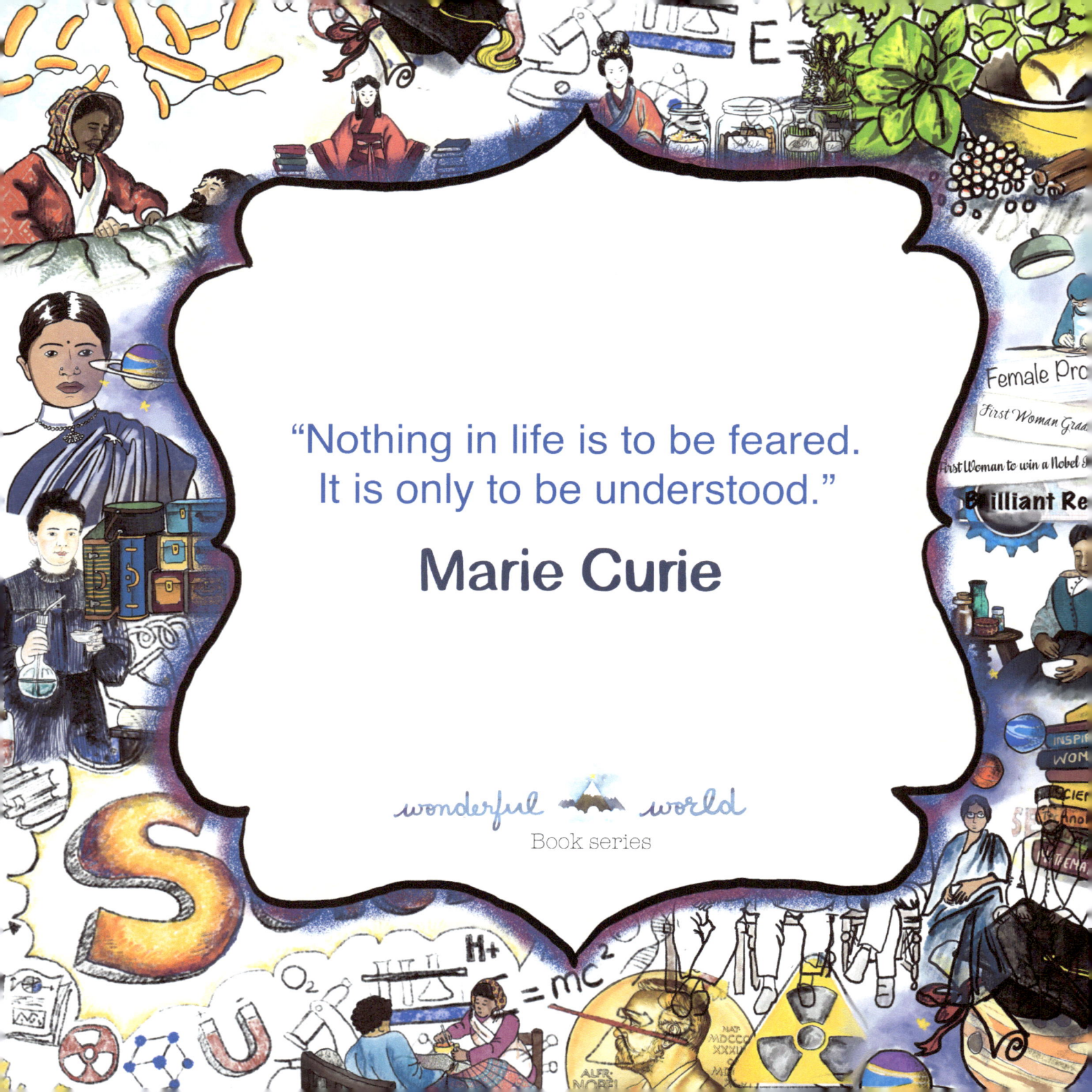

"Nothing in life is to be feared. It is only to be understood."

Marie Curie

wonderful world
Book series

Even when
Marie Curie was small,

She loved

the most of all.

She wanted to be a

scientist who would

Scientist: A person who knows a lot about Science.

Use **Science** to do a

lot of good.

She worked very hard
so when she grew,

She became a **Physicist** and a **Chemist** too.

Physicist: A scientist who has studied Physics —
the study of how things move and what makes them move.
Chemist: A scientist who has studied Chemistry —
the study of what things are made of and how they work.

Marie and her husband

Pierre Curie,

developed the

Theory of Radioactivity,

Theory: An explanation of why things happen or how things work.
Radioactivity: When small units that make up an object (thing) release energy (which is usually not safe).

For which they won
a prize called the

Nobel,

Along with
Henri Becquerel.

Marie got a second Nobel prize

for finding the elements,

Element: Purest form of a substance (thing) like gold, silver, oxygen etc.
There are 118 elements which combine to form other things.

Radium and Polonium

after many experiments.

Experiments: A scientific test done to see if an idea or theory is correct.

She was the first person

to win two,

A wonderful person —
just like you!

Glossary

Scientist
A person who knows a lot about Science.

Physicist
A scientist who has studied Physics - the study of how things move and what makes them move.

Chemist
A scientist who has studied Chemistry - the study of what things are made of and how they work.

Theory
An explanation of why things happen or how things work.

Glossary

Radioactivity
When small units that make up an object (thing) release energy (which is usually not safe).

Element
Purest form of a substance (thing) like gold, silver, oxygen etc. There are 118 elements which combine to form other things.

Experiment
A scientific test done to see if an idea or theory is correct.

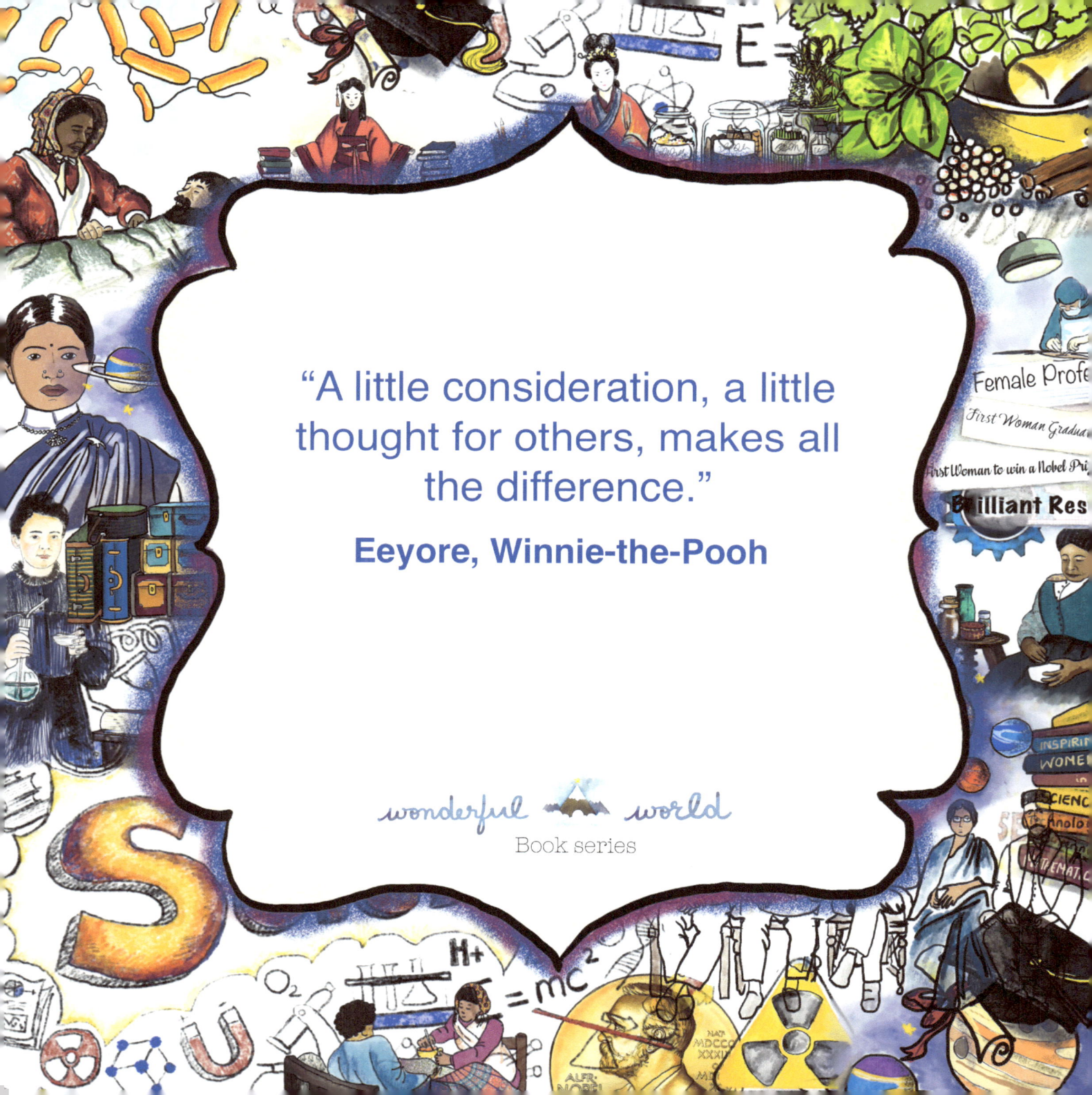

"A little consideration, a little thought for others, makes all the difference."

Eeyore, Winnie-the-Pooh

wonderful world
Book series

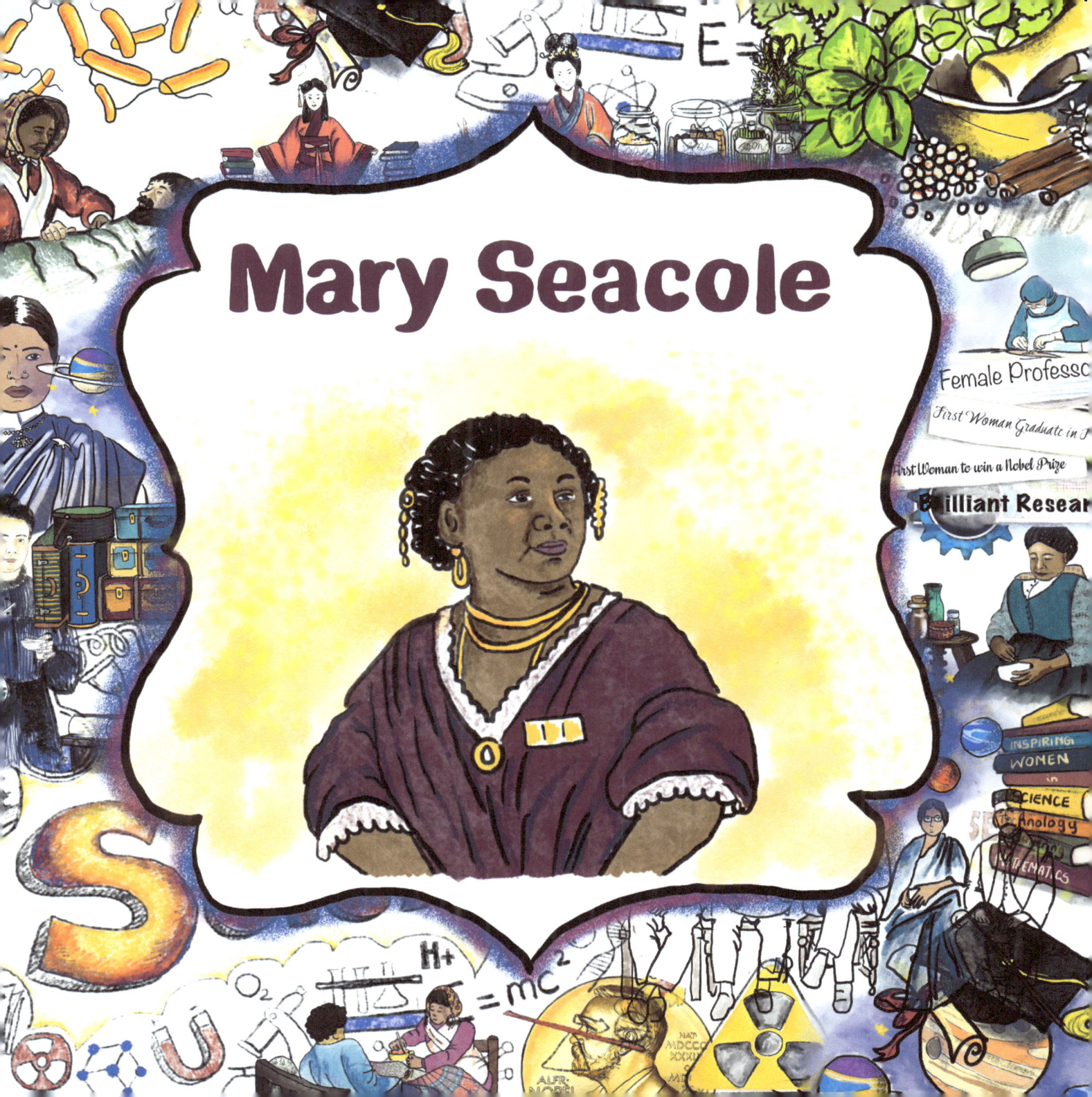

Ever since she was a **little girl,**

Mary Seacole
wanted to travel the world.

She was a
nurse and a doctress,

Nurse: A person who works in a hospital and looks after people who are sick or hurt.

Doctress: A woman doctor who makes medicines from plants to cure people.

Who was known for her **kindness.**

She made pills from

plants and **trees**

To cure every type of

disease.

Disease: An illness - when something bad happens to someone's mind or body.

When cholera made
people very ill,

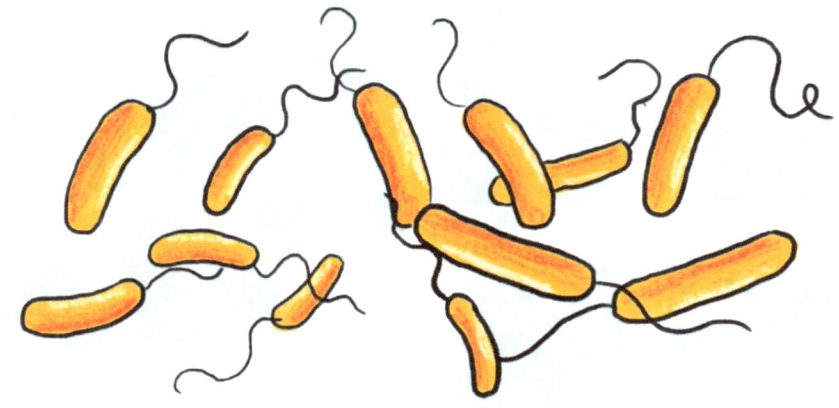

She nursed them to health with tender skill.

During the time of war

in Crimea,

Crimea: A place in Southern Ukraine in Europe.
It is surrounded by the Black Sea on three sides.

She travelled from Jamaica

with a great idea.

Jamaica: An island in the Caribbean Sea, in North America.

She set up a house called the
'British Hotel'

Where hurt soldiers

stayed until they got well.

She was **hardworking,**

brave, and **resourceful** too,

Resourceful: A person who handles difficult situations well.

A wonderful person —
just like you!

Glossary

Nurse
A person who works in a hospital and looks after people who are sick or hurt.

Doctress
A woman doctor who makes medicines from plants to cure people.

Disease
An illness - when something bad happens to someone's mind or body.

War
A fight between big groups of people or between countries.

Glossary

Crimea
A place in Southern Ukraine in Europe. It is surrounded by the Black Sea on three sides.

Jamaica
An island in the Caribbean Sea, in North America.

Resourceful
A person who handles difficult situations well.

"We are responsible for the evils we tolerate, and for those we do not resist actively."

Dr. Muthulakshmi Reddy

wonderful world
Book series

Once upon a time, there lived

Muthulakshmi
(moo-thoo-luck-sh-me)

Who was meant to be a

Devadasi.
(they-vuh-dha-see)

Devadasi: A girl who had to live in the temple and do everything the village elders and temple priests told her.

Devadasis were girls who danced in the

temple yard.

Temple: A place where people go to pray.

Instead, her father told her to

study hard.

She had to get the
Maharaja
(muh-ha-raa-jaa)

to agree,

Maharaja: A word for 'King' in many Asian languages.

So she could go to

college

to get a degree.

She worked very hard

so when she grew,

She became a **doctor**

and a **reformer** too.

Doctor: A person who cures people of their illness.
Reformer: A person who works to bring good change in the world around us.

One of the first women to

study **surgery,**

Surgery: The doctor uses tools on the body or inside it to fix something that is wrong or to treat illness.

She set other Devadasi girls free,

So none of them were forced

to be **dancers.**

And she started a hospital

to cure different **cancers.**

Cancer: A type of illness when a very tiny part called a cell grows really fast and spreads inside the body.

She was **brave**, **kind**,

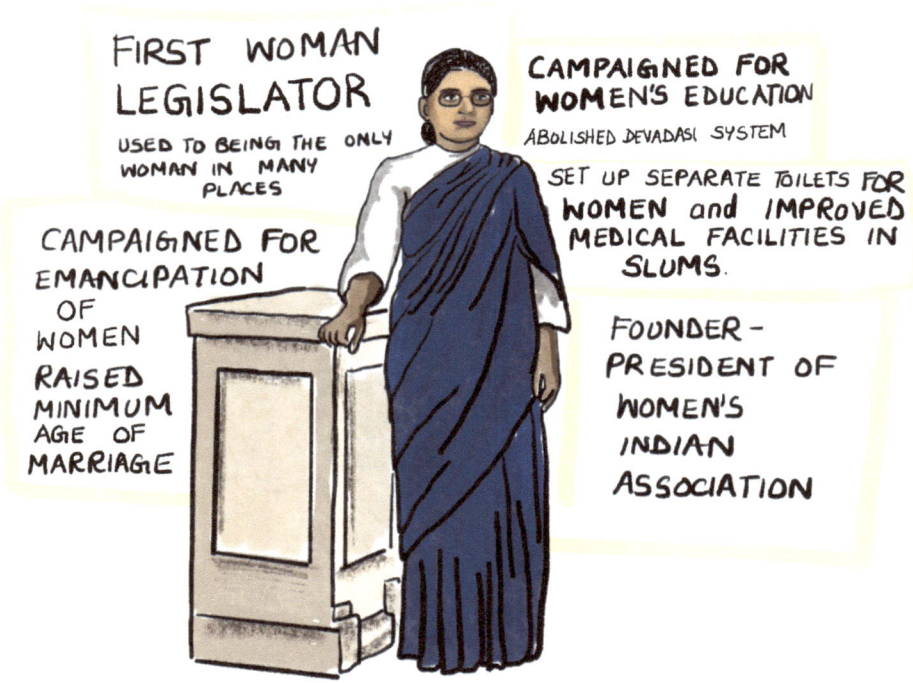

and **clever** too,

A wonderful person —
just like you!

Glossary

Devadasi
A girl who had to live in the temple and do everything the village elders and temple priests told her.

Temple
A place where people go to pray.

Maharaja
A word for 'King' in many Asian languages.

Doctor
A person who cures people of their illness.

Glossary

Surgery
The doctor uses tools on the body or inside it to fix something that is wrong or to treat illness.

Cancer
A type of illness when a very tiny part called a cell grows really fast and spreads inside the body.

Reformer
A person who works to bring good change in the world around us.

"If you can dream it, you can do it."

Walt Disney

wonderful world
Book series

Wang Zhenyi

Once upon a time, during the **Chinese dynasty** of **Qing,**
(Ching)

Dynasty: An era (many years) during which all the kings and queens were from the same family.

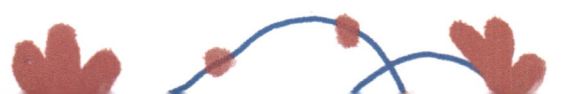

lived **Wang Zhenyi**
(Waa-ng Zh-en-yee)

who loved studying.

She studied the **sun**, the **moon**, the **stars**,

And **Jupiter, Saturn, Venus,** and **Mars.**

She loved **numbers,**

being a

mathematician,

Mathematician: A person who is an expert in Mathematics (the study of numbers).

And wrote a book,

'The Musts of Calculation'.

Though learning was hard,

she did not quit.

She worked and practised

until she was good at it.

She traveled all over
the countryside,

And learnt to aim, fight, hunt, and ride.

She believed everyone must work hard so,

They can be anything they wish when they grow.

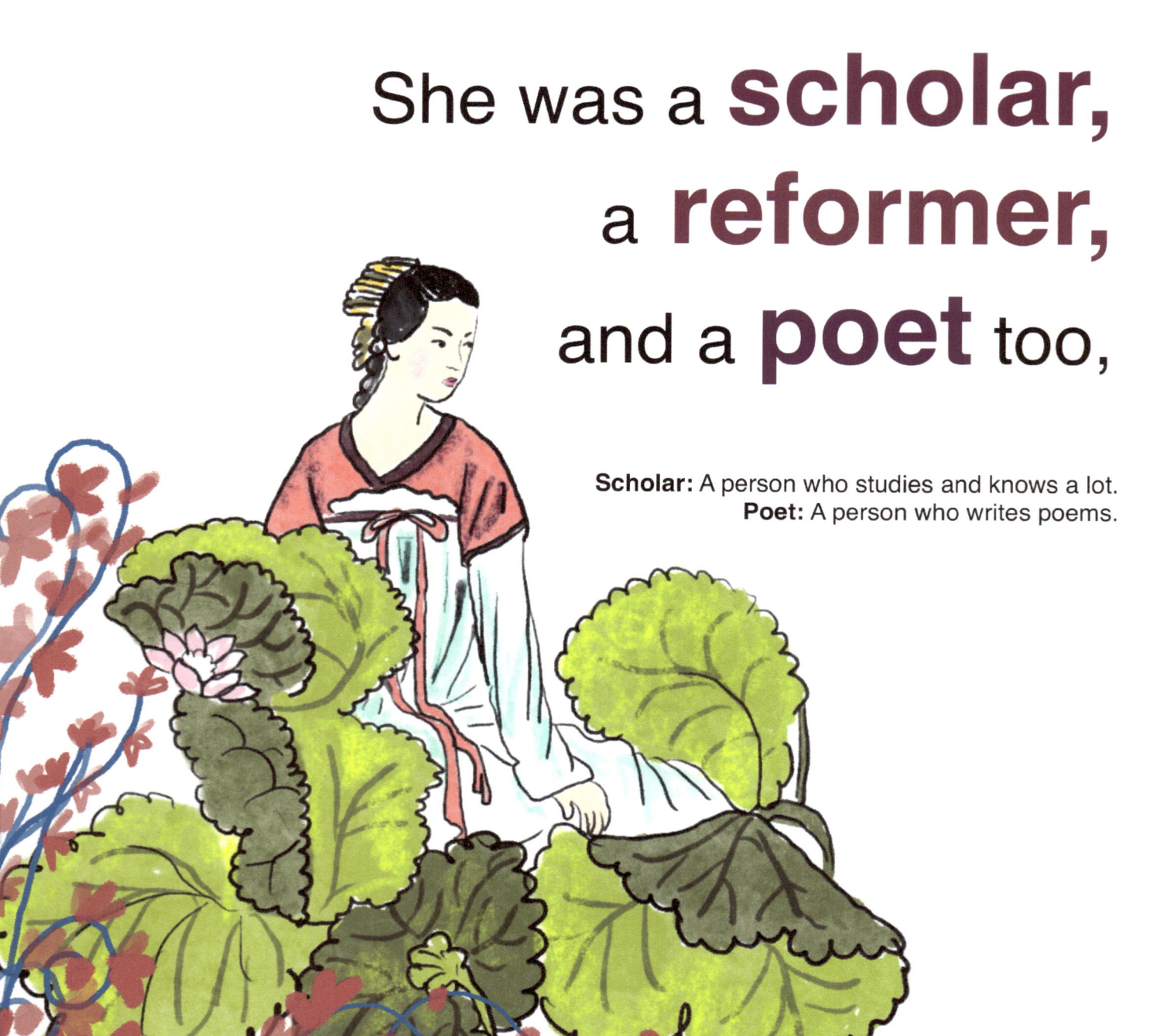

She was a **scholar**, a **reformer**, and a **poet** too,

Scholar: A person who studies and knows a lot.
Poet: A person who writes poems.

A wonderful person — just like you!

Glossary

Dynasty
An era (many years) during which all the kings and queens were from the same family.

Mathematician
A person who is an expert in Mathematics (the study of numbers).

Scholar
A person who studies and knows a lot.

Poet
A person who writes poems.

wonderful world
Book series

The Beginning

Thanks for reading my book.
I hope you've enjoyed it. For an independent author, ratings are very important for the success of their book. I'd be grateful if you could take a minute to rate this book on Amazon/ Goodreads.
Your support makes all the difference.

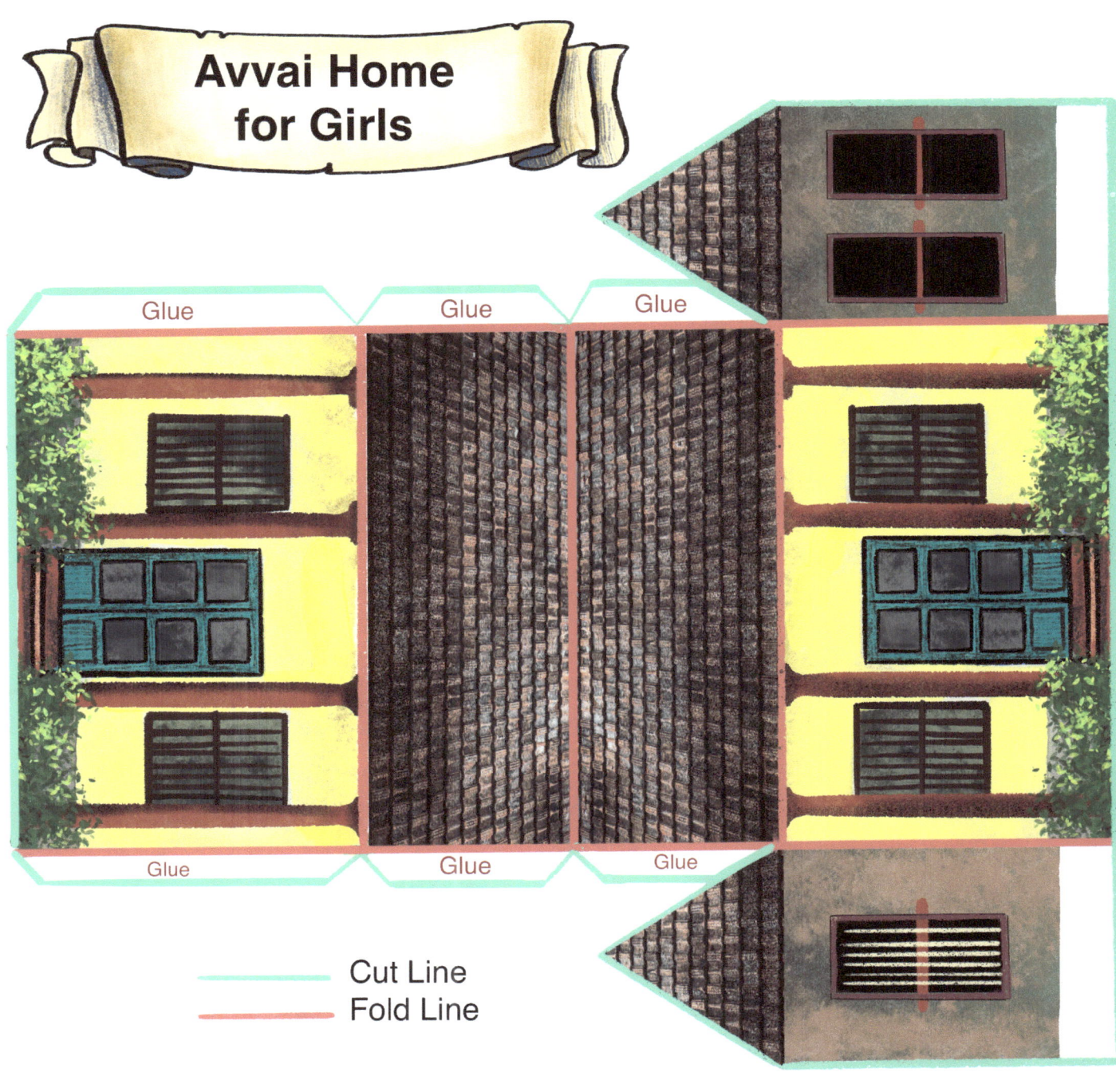

Cut Line
Fold Line

The British Hotel

wonderful world
Book series

About the Author

Author, illustrator, and dentist, **Ramya Julian** finished her first novel at the age of ten and she avers it was very well received though it was read only by her brother.

She has all the hobbies of a maiden Victorian aunt – reading, writing, painting, crocheting, knitting and sewing, and the temperament of one. When she's not guilt-tripping her two daughters into good behaviour, she can be found devouring books, crafting poems and puns, and chuckling at her own witticisms. She grew up in India and now lives with her husband and their two daughters in London.

She has experienced so much joy through the enchanting artistry of many authors and creators, that she aspires to share at least some of it through her writing.

To see more of her work, visit **www.ramyajulian.com**

www.ramyajulian.com

Also in this series

NEXT IN LINE: MANY MANY MORE WONDERFUL DIVERSE HEROES

TO MY NEWSLETTER
For the latest news and free printables
www.ramyajulian.com

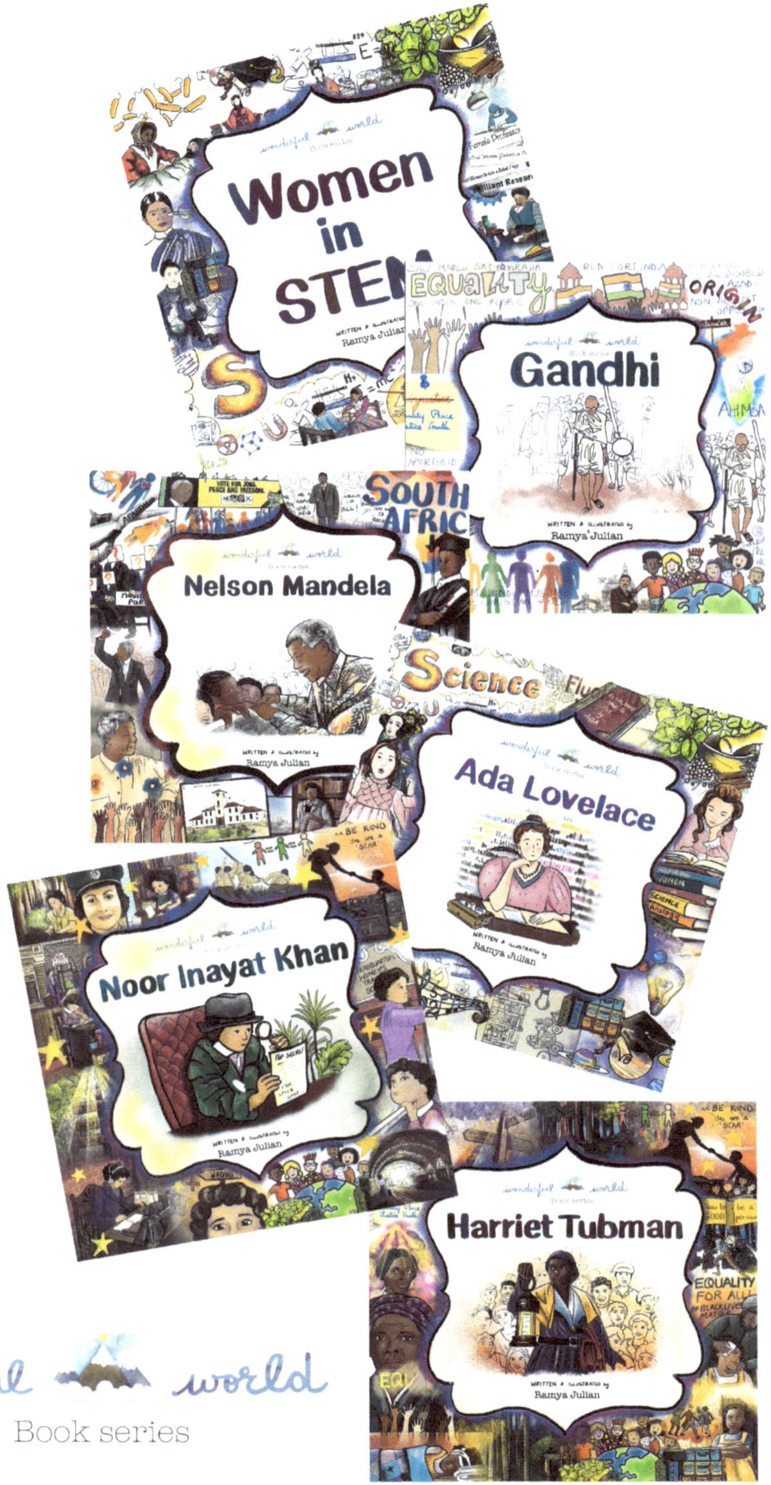

@RAMYAJULIAN

Book series

www.ingramcontent.com/pod-product-compliance
Lightning Source LLC
Chambersburg PA
CBHW050748110526
44591CB00002B/12